For Mum and Dad

Expert consultant: Stuart Atkinson

First published 2017 by Walker Studio, an imprint of Walker Books Ltd, 87 Vauxhall Walk, London SE11 5HJ
This edition published 2018

2 4 6 8 10 9 7 5 3 1

This book had been typeset in Nevis

Printed in China

British Library Cataloguing in Publication Data: a catalogue record for this book is available from the British Library

ISBN 978-1-4063-8715-5

www.walkerstudio.com

CURIOSITY

The Story of a Mars Rover

MARKUS MOTUM

Wherever you are in the world right now,
I'm a very long way away.
I'm not even on the same planet as you.

I'm a Mars rover. A rover is a moving robot, built to explore far-off places – places too far away or dangerous for humans to visit.

How did I get here? Why was I sent?

This is my story.

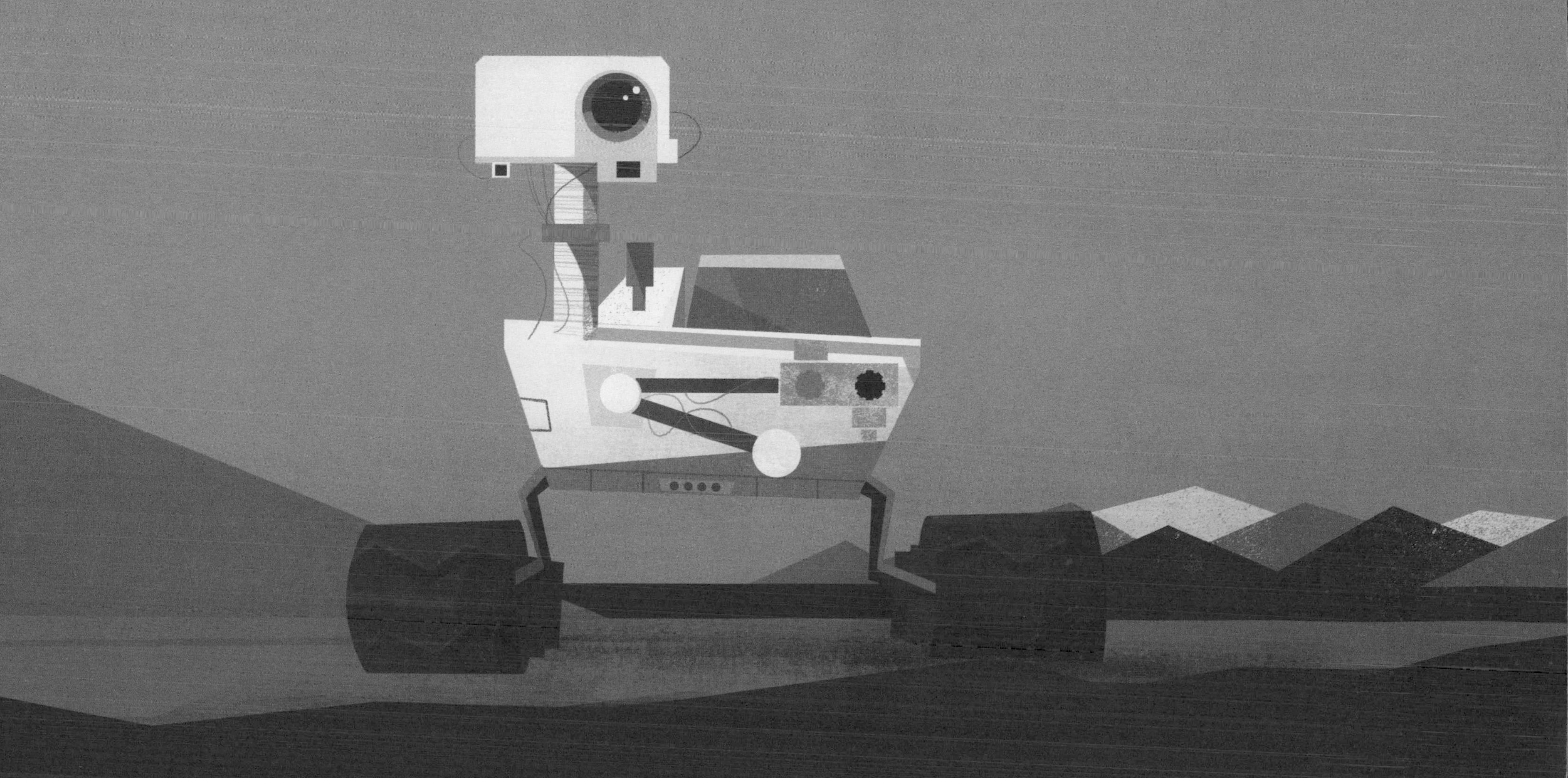

Since the beginning of recorded history, humans have wondered about Earth and its place in the universe. Although scientists have made great discoveries and explained some of the mysteries of space, many questions remain unanswered. The planets hold their secrets well. Of these, one question intrigues humans above all others:

Is there anybody else out there?

Scientists decided that the best place to look for other life was on one of our closest neighbours...

MARS

The red planet.

As well as being one of the closest, Mars is the planet in our solar system most similar to Earth. Scientists think that in the past, Mars had environments similar to the ones that Earth has today. Mars is covered in red dust now, but there is evidence that millions of years ago the planet had lakes, flowing rivers and even great oceans.

Why does this matter?

Because everywhere on Earth there is water, there is life!
However, there is one problem...

The trip to Mars in a rocket can be over

350,0

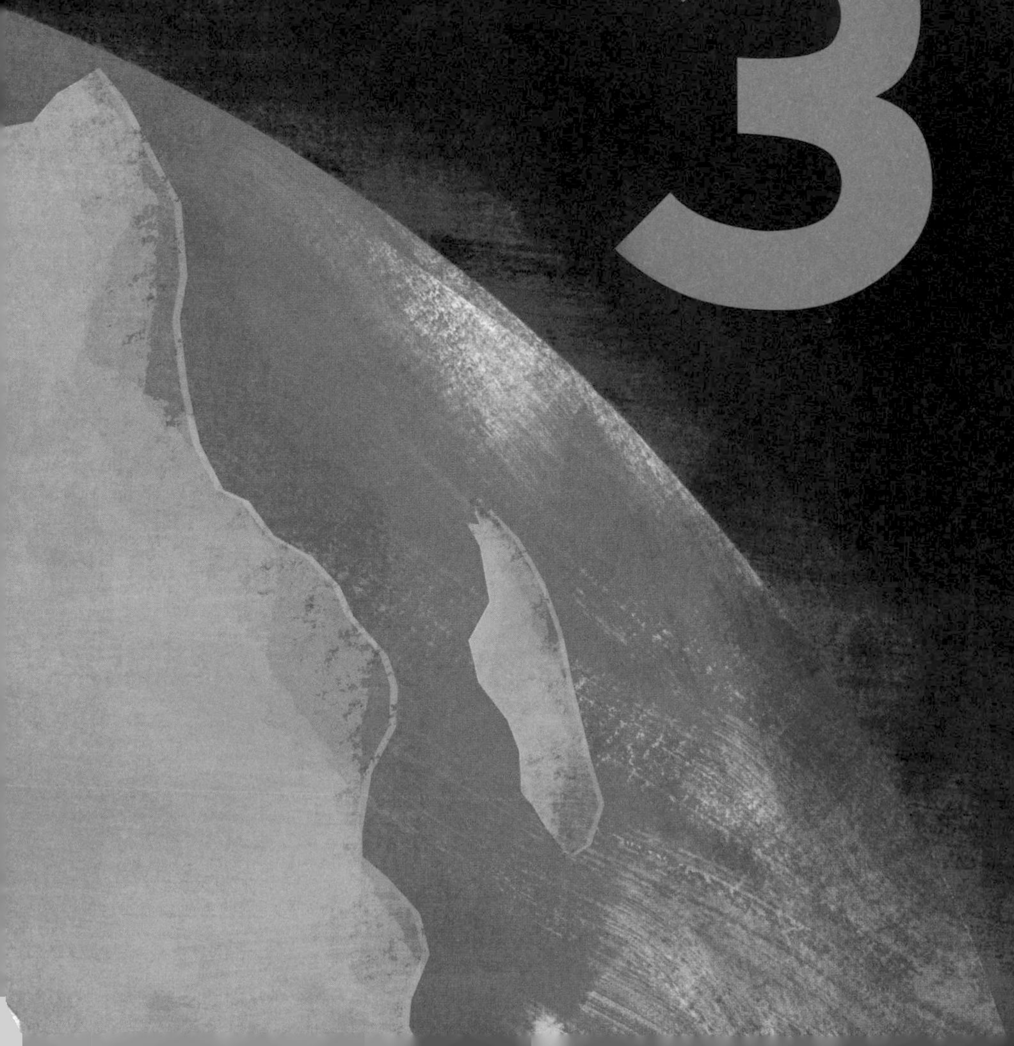

00,000

miles long.

That's three hundred and fifty million miles long,

a distance much, much further than any human has ever travelled in space.

Even with our modern technology, we don't have a practical way to get humans to Mars – the journey would take at least six months – or to get them back.

Humans have been able to explore closer to home. On 20 July 1969, the Apollo 11 mission landed: an astronaut took one famous step and became the first person on the Moon. For the few hours Neil Armstrong and Buzz Aldrin walked on its surface, people could look up at the Moon and know that someone might be looking back at them.

Mars is not only about 200 times further away than the Moon, a journey humans would find difficult to undertake, but it is also an inhospitable environment. But what if it was possible to send an explorer into space who didn't need food, water or oxygen? That is where I come in.

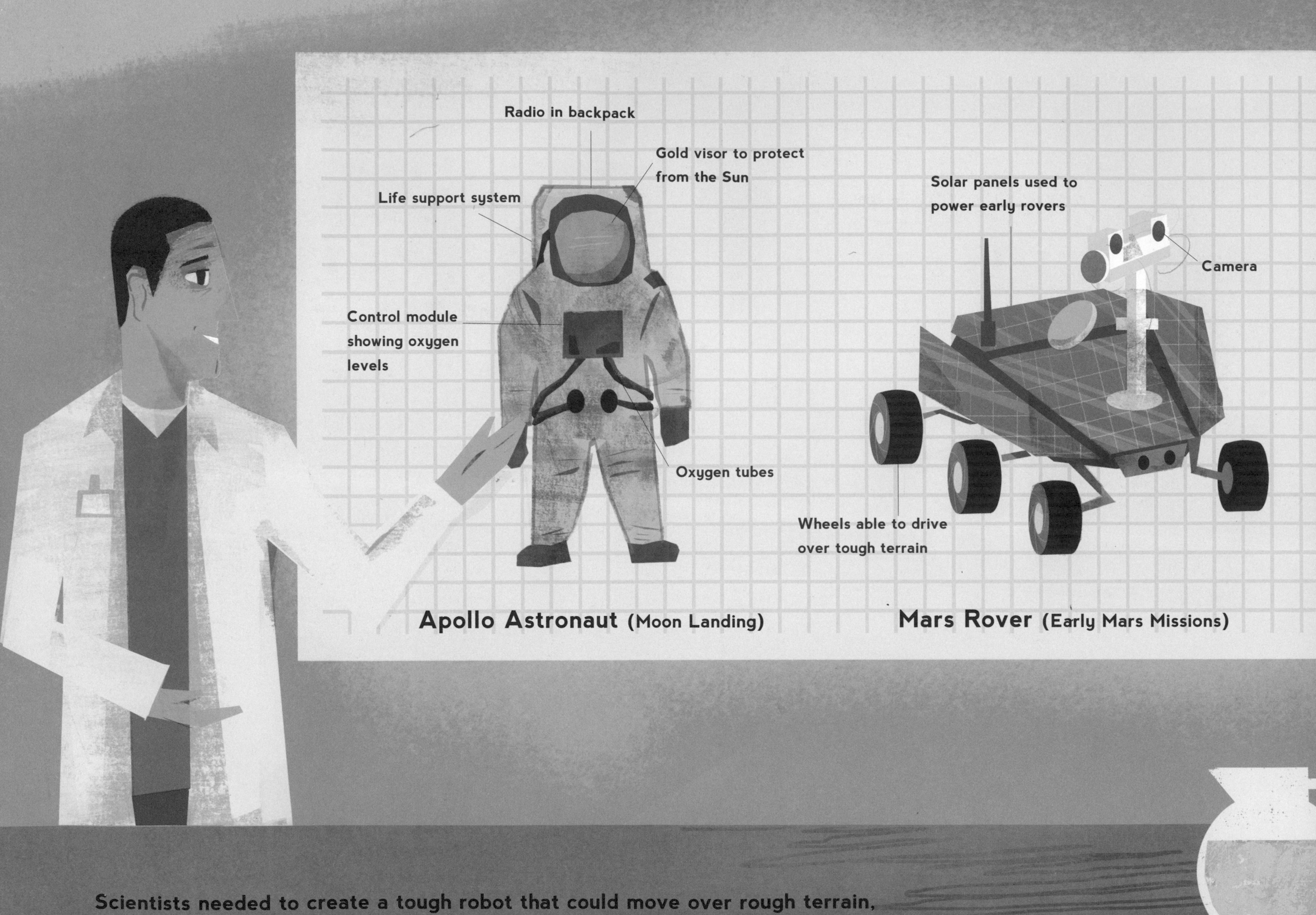

Scientists needed to create a tough robot that could move over rough terrain, with all the skill and equipment a scientist would need to investigate Mars.

So, the National Aeronautics and Space Administration (NASA) had the idea to build rovers like me.

Deimos
(moon)
Mars
Phobos
(moon)

The project to build me began at the Jet Propulsion Laboratory near Los Angeles, California. I needed to be larger and more advanced than any of the rovers NASA had built before.

Previous successful rovers had taken photos of Mars, giving us never before seen images. Other missions were "orbiters" and "fly-bys" – flying around or past Mars to gather information without even landing on the planet.

However, Mars is a dangerous and tricky destination. By 2007, 39 missions had been launched to the red planet and over half had ended in failure. Some of these earlier missions had become lost in space, while others had crashed into Mars, never to be heard from again.

The Jet Propulsion Lab where I was built had to be kept as clean as possible. Everyone had to wear white overalls, which were nicknamed bunny suits. These suits stopped germs from spreading to me and my equipment.

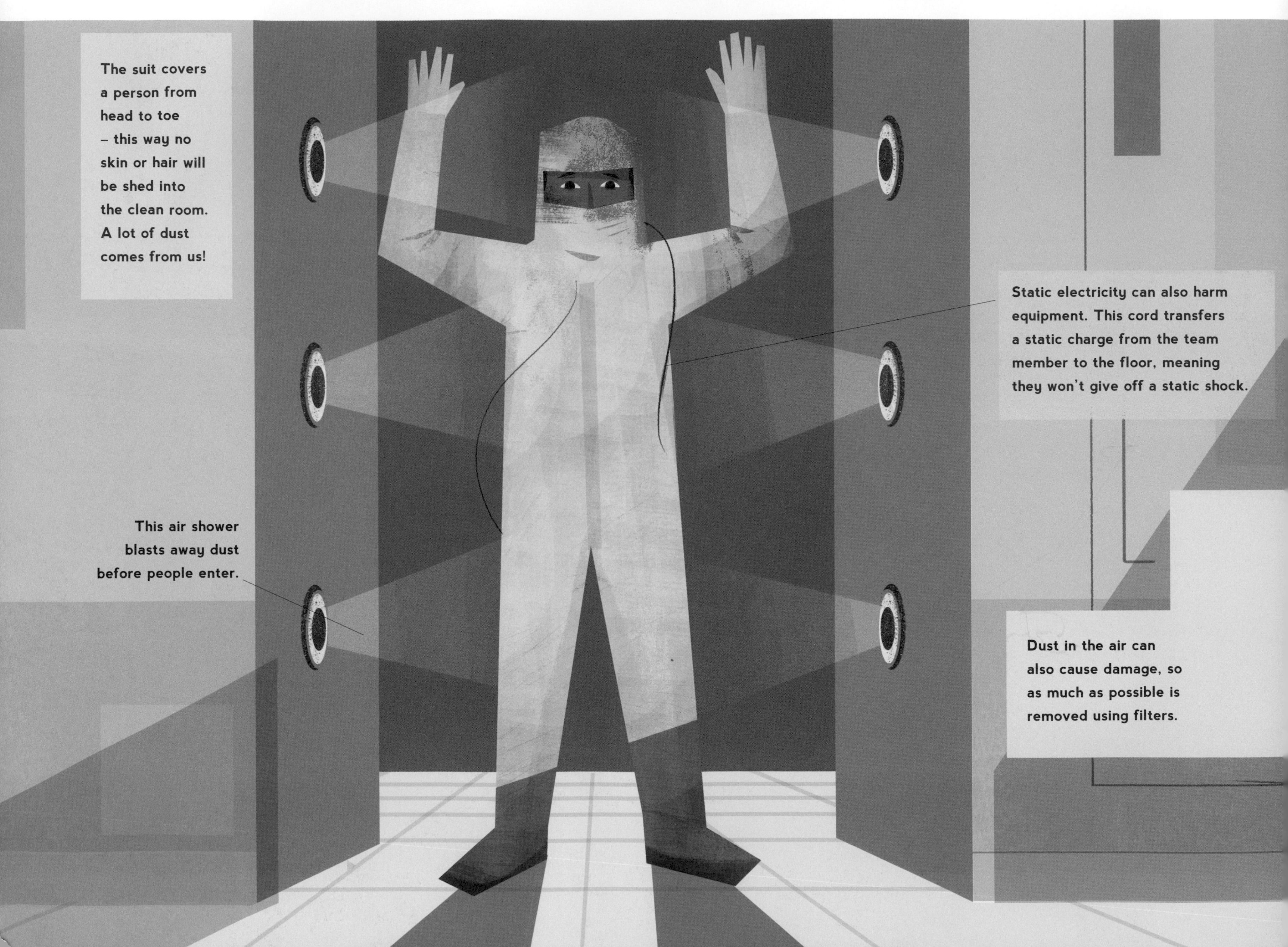

The last thing the team wanted was to think they had discovered tiny forms of life on Mars, only to realize bacteria had come with me from Earth.

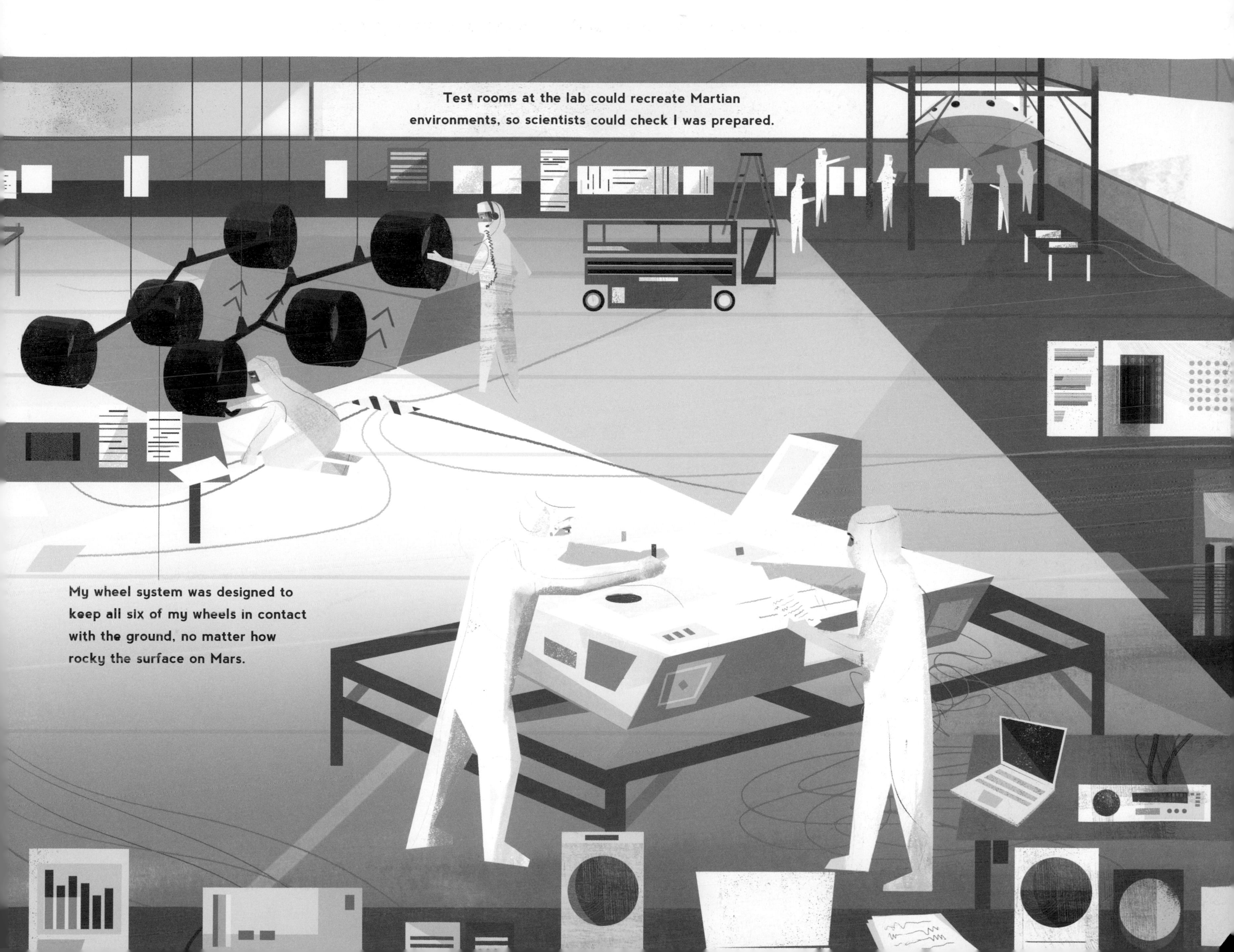

Entire new technologies had to be invented for my mission. I needed to be able to carry a lot of equipment to test what I found on Mars – literally my own laboratory. To give the mission as good a chance of success as possible, years of testing were needed to make sure everything would work correctly first time. After all, if something goes wrong on Mars, no one can come and fix me.

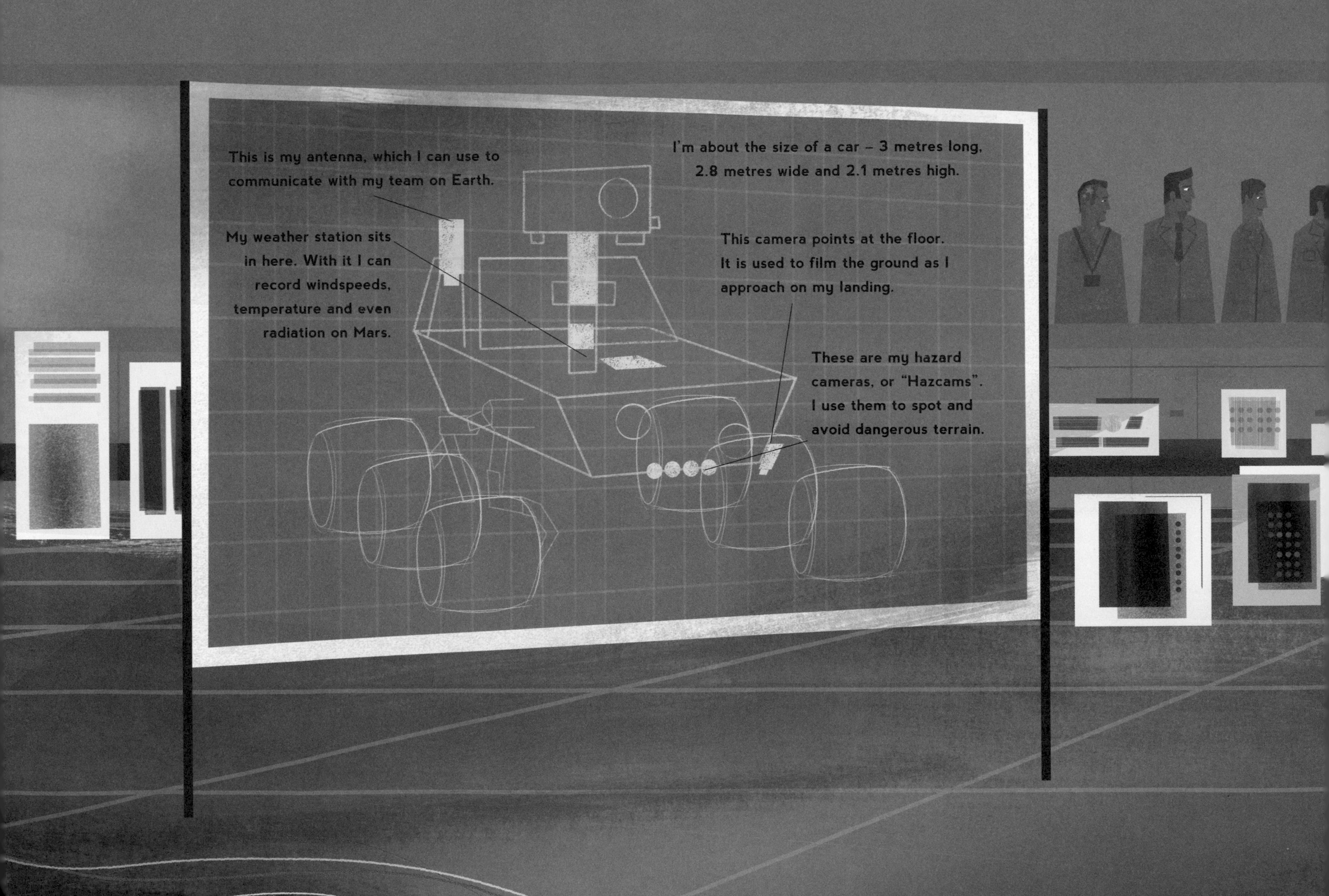

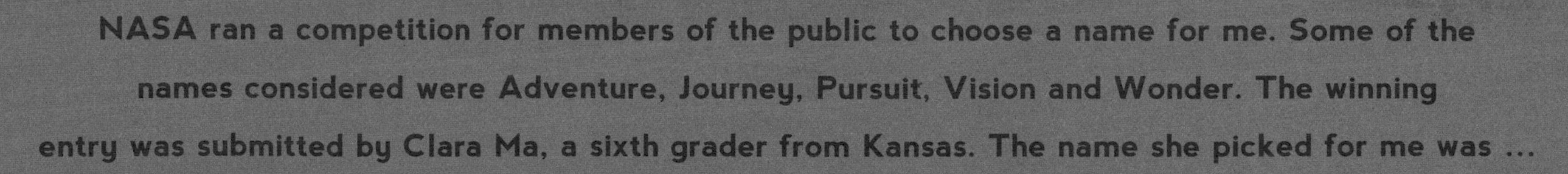

NASA ran a competition for members of the public to choose a name for me. Some of the names considered were Adventure, Journey, Pursuit, Vision and Wonder. The winning entry was submitted by Clara Ma, a sixth grader from Kansas. The name she picked for me was ...

CURIOSITY

My body carries the on-board chemistry lab where I run all my tests.

Just like you I have a shoulder, elbow and wrist so I can be as flexible as possible.

This is my battery. It's nuclear powered. The plutonium in here will power me for years.

My robotic hand has many tools, including a drill and a brush to see what's under all the dust on Mars.

These wheel treads help me judge how far I've moved by measuring the marks they leave.

Now I was ready to fly to Mars. But I couldn't take off from California, where I was built. When I launched, my flight into space would take me up and eastwards. This meant it was safest to launch where the ocean was to the east, just in case something went wrong. I would have to fly to Florida to begin my space mission.
So I boarded a carrier plane ...

and flew from Los Angeles ...

to the Kennedy
Space Center.

The rocket that would take me into space was waiting for me in Florida. The Atlas V rocket is around 200 feet tall – the same as a nineteen storey building – and it is almost all rocket fuel. This is because it needs to be powerful enough to launch whatever it is carrying, known as its payload, into space.

The Atlas V is known as an expendable launch vehicle, which means its parts can only be used once. I was placed inside the nose cone at the very top. The cone is like a shield that protects the rest of the rocket as it breaks through the Earth's atmosphere.

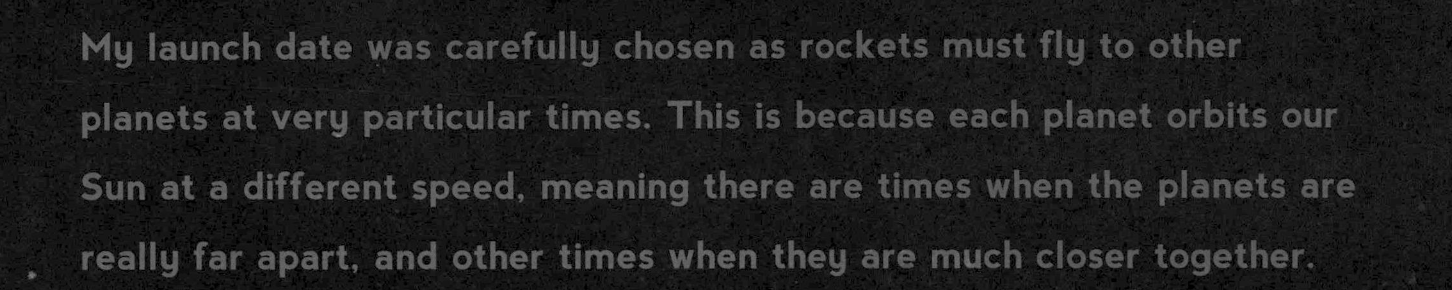

My launch date was carefully chosen as rockets must fly to other planets at very particular times. This is because each planet orbits our Sun at a different speed, meaning there are times when the planets are really far apart, and other times when they are much closer together.

Every few years, Earth and Mars are lined up so the distance between them allows missions to reach Mars using less fuel.

If we missed this opportunity the mission would have been delayed for years.

Launch day arrived – 26 November 2011. I was ready to fly to Mars. In the control room, the team ran through the final checks.

At last the rocket was ready for countdown:

Five, Four, Three, Two, One –

Blast Off!

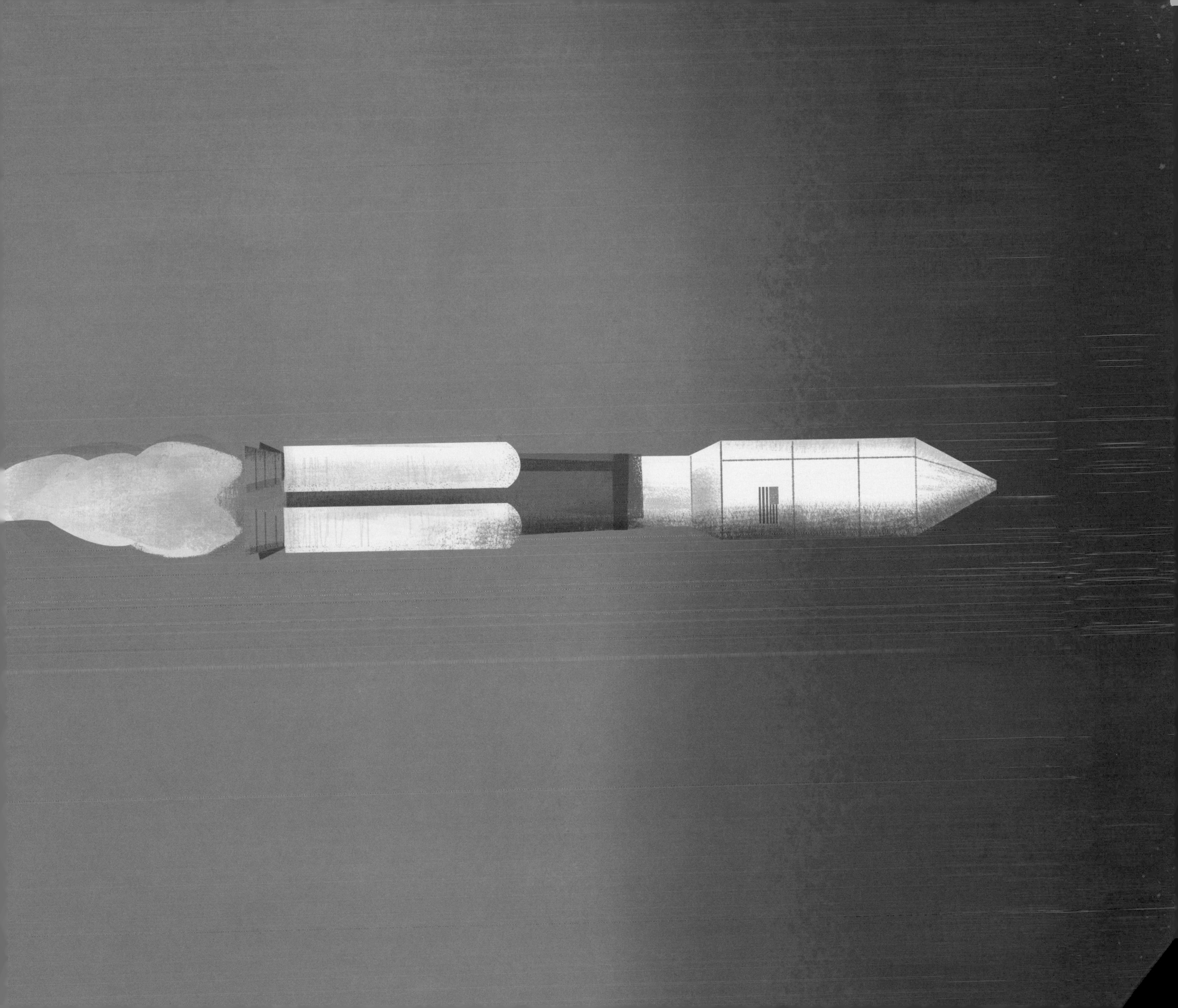

The first part of my journey was breaking through Earth's atmosphere and reaching space. Once the fuel had done its job of getting the rocket into space, stage-by-stage the empty boosters broke off and fell safely into the ocean.

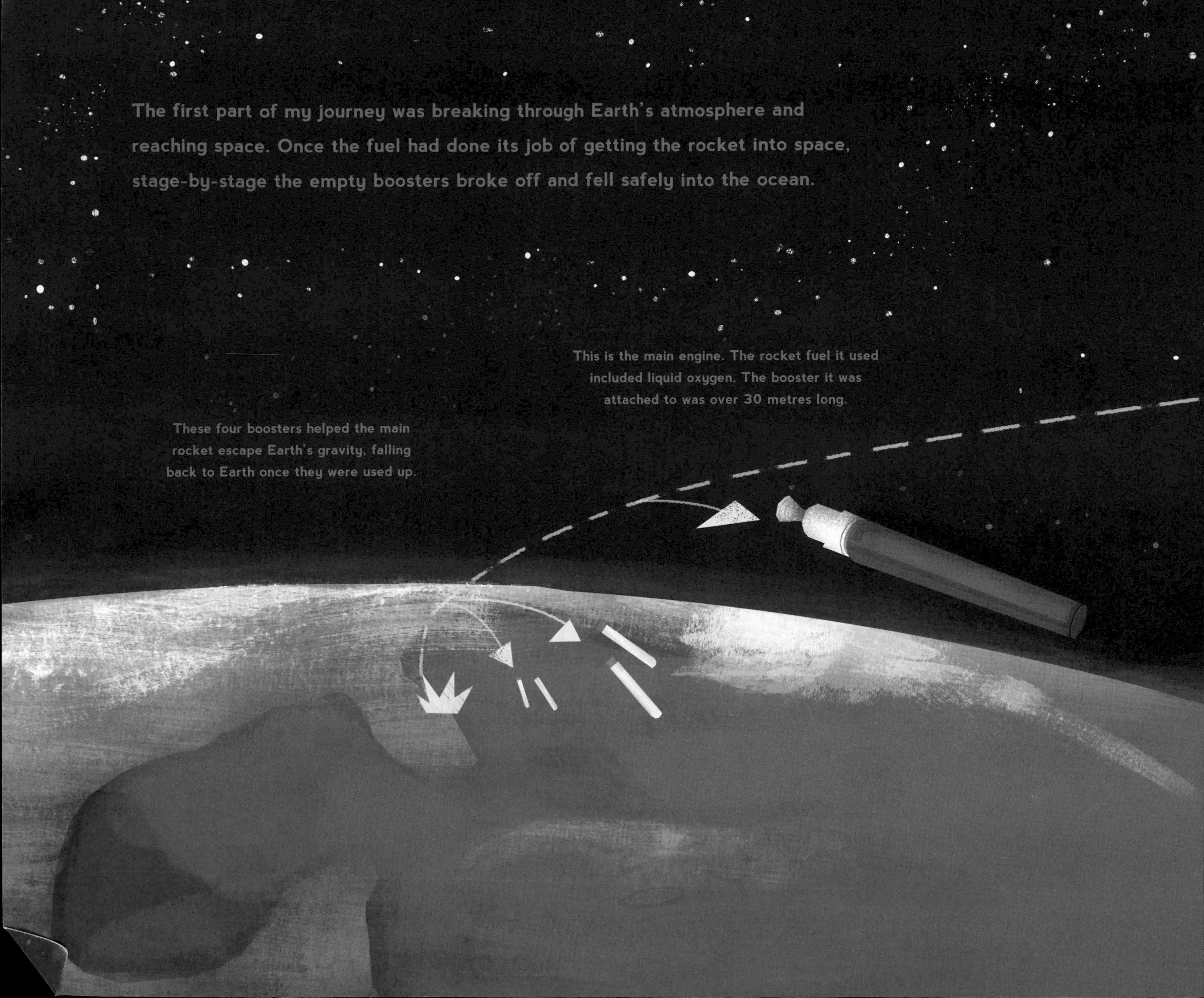

By now the rocket had carried me safely into space. At this stage the nose cone protecting me broke off and fell back towards Earth.
Forty minutes after take-off, once I pushed through Earth's atmosphere, the final rocket booster detached.
I travelled to Mars in this protective shell called the module. Inside the module a star scanner and Sun sensor directed the journey to Mars. Thrusters on the outside of the module changed direction if necessary. It was the only part of the rocket that made the long journey to Mars with me.

After 253 days

of hurtling through space, I finally arrived on Mars, travelling at 13,000 miles per hour! Next would be the trickiest part of my entire journey – landing safely.

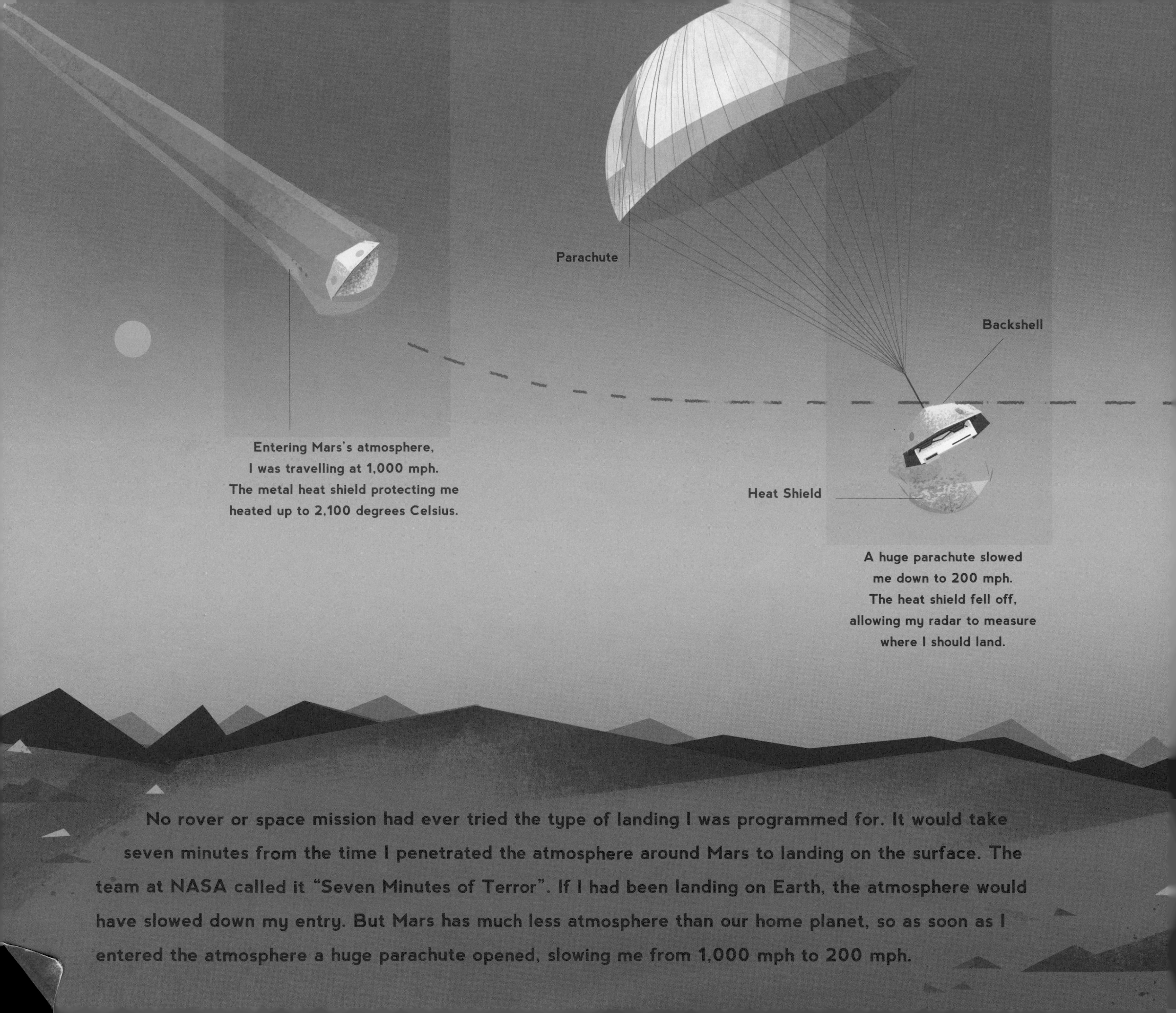

No rover or space mission had ever tried the type of landing I was programmed for. It would take seven minutes from the time I penetrated the atmosphere around Mars to landing on the surface. The team at NASA called it "Seven Minutes of Terror". If I had been landing on Earth, the atmosphere would have slowed down my entry. But Mars has much less atmosphere than our home planet, so as soon as I entered the atmosphere a huge parachute opened, slowing me from 1,000 mph to 200 mph.

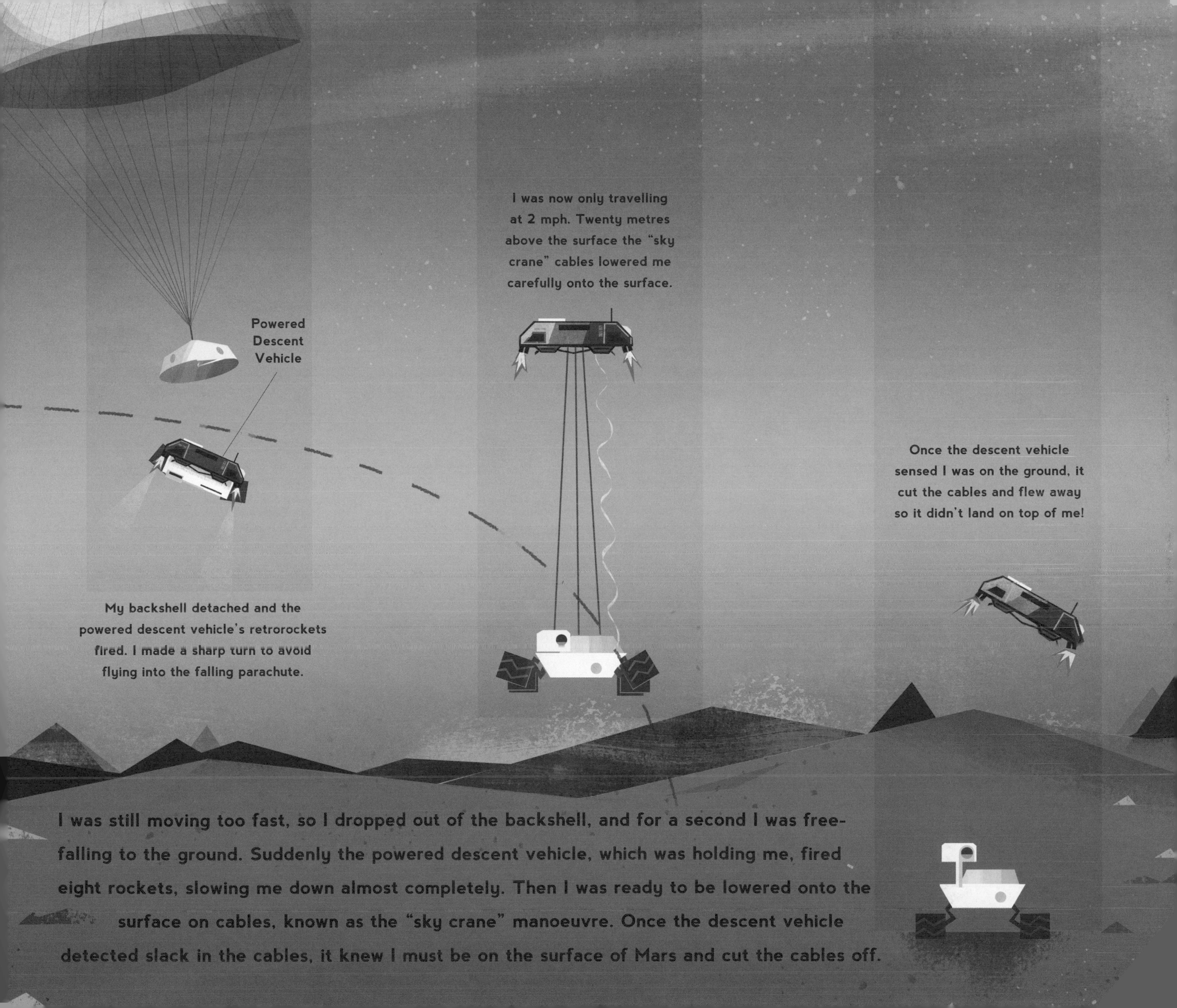

I was still moving too fast, so I dropped out of the backshell, and for a second I was free-falling to the ground. Suddenly the powered descent vehicle, which was holding me, fired eight rockets, slowing me down almost completely. Then I was ready to be lowered onto the surface on cables, known as the "sky crane" manoeuvre. Once the descent vehicle detected slack in the cables, it knew I must be on the surface of Mars and cut the cables off.

I made it.

As soon as I was safely on the surface, I sent a message to my NASA team letting them know I had landed. Because of the huge distance between Mars and Earth, the message took fifteen minutes to get back to Earth. It was a tense time in the control room as everyone waited. Then the words everyone wanted to hear were read out:

"Touchdown confirmed. We're safe on Mars!"

NYC
CH 3 NEWS
BREAKING NEWS
CURIOSITY HAS LANDED!

It wasn't just the NASA team who watched my landing. All around the world, people tuned in to see me safely arrive after my long journey. Crowds gathered in Times Square, New York to watch on a giant screen. It was an exciting but scary time. After all, if just one thing went wrong, all contact between me and the team back on Earth would be lost.

When they heard I'd landed some people cheered – others breathed a huge sigh of relief!

The first thing I did on Mars was to send images of myself back to NASA to make sure I hadn't been damaged during the journey. My landing had been perfect – if a little dusty. The site had been carefully selected by scientists as a place likely to have evidence of water. Now it was time to start looking, but where?

The NASA team looked at Mars through my 17 cameras and chose interesting-looking spots on the surface. Once something was identified I moved towards it. I can move about 200 metres per day, with my wheel tracks measuring the distance I travel.

NASA wants to find out what is under the surface of Mars to understand the history of the planet and how it was formed. The deeper I dig the more information I can gather.

I am able to drill into rocks and scoop up material. I then test it in my on-board laboratory. Gradually, by piecing together information from different locations, NASA hopes to build a picture of the planet's past and perhaps discover why Mars changed from being a warm planet with water to the cold, red planet it is today.

I take photos from my navigation cameras and my hazard cameras, and combine what I see in both pictures to find a safe way to drive around Mars.

Firing the laser at these rocks lets us find out what they are made of, which means finding out what Mars used to be like as a planet.

As a hole is drilled, the powder gets sent up a tube into my arm. My arm then reaches over and places the powder into my laboratory for testing.

There are many questions still to be answered. What was the Mars environment like long ago? How suitable was it for life? Did we once have neighbours on Mars? The tests I carry out will help provide as many answers as possible. Most likely, I will find something which leads to more questions!

Luckily, with space exploration, questions can be just as exciting as answers. Thanks to the curiosity of explorers, Neil Armstrong's footprints are on the Moon. And now, my tyre tracks are being left on another planet. Perhaps one day soon, footprints from the next generation of explorers will join mine.

“Curiosity is the passion that drives us through our everyday lives. We have become explorers and scientists with our need to ask questions and to wonder. We will never know everything there is to know, but with our burning curiosity, we have learned so much.” Clara Ma

MARS ROVERS

Although Curiosity is the most advanced rover sent to Mars so far, she is not the only rover to explore the red planet. In 1997, the Sojourner rover was the first to land successfully on Mars, and remained active for much longer than expected; it was nearly three months before Sojourner sent her final transmission to Earth.

The Mars Exploration Rover Spirit was the next to make a successful landing in 2004. She also survived much longer than she was expected to and discovered incredible evidence that there was once water on Mars. After travelling nearly five miles over five years, Spirit got stuck in dust and eventually lost contact with Earth.

But Spirit's twin rover is still operating today. Opportunity landed three weeks after Spirit, and since then has travelled further than the length of a marathon! No other robot has ever gone so far on the surface of another world. Opportunity continues to explore Mars's craters and send back images, many years after she was expected to shut down.

Where will our curiosity take us next?

NASA MARS ROVER MISSION OBJECTIVES:

- Determine whether life ever arose on Mars
- Characterize the climate of Mars
- Characterize the geology of Mars
- Prepare for human exploration

TIMELINE OF MARS MISSIONS

1964–5
Mariner 4 makes first successful fly-by of Mars and sends back close-up images

1969
Mariner 6 and Mariner 7 fly-bys

1971–2
Mariner 9 orbiter

1975–82
Viking 1 orbiter and lander

1975–87
Viking 2 orbiter and lander

1996–7
Mars Pathfinder lander and Sojourner rover

1996–2006
Mars Global Surveyor orbiter

2001–present
Mars Odyssey orbiter

2003–present
Mars Express orbiter (Beagle 2 lander lost on arrival)

GLOSSARY

APOLLO 11 – the spaceflight that landed the first two humans (the Americans Neil Armstrong and Buzz Aldrin) on the Moon on 20 July, 1969

ATMOSPHERE – the gases that surround a planet

BACTERIA – single-celled organisms that live on, in and around most living and non-living things

BOOSTERS – the first stage of a rocket or spacecraft that provides acceleration before they are jettisoned

GRAVITY – the force that pulls things towards the centre of Earth and other planets

LIFE SUPPORT SYSTEM – a pack that provides astronauts with oxygen while removing carbon dioxide. It also typically includes water-cooling equipment, a fan and a radio.

ORBIT – the repeated course of an object around a star, planet, asteroid, comet or galaxy

POWERED DESCENT VEHICLE – vehicle that fires retrorockets to control and slow landing of a rover

RADAR – a method that sends out radio waves to detect the distance and location of objects

SOLAR SYSTEM – a group of planets orbiting a sun. Our solar system has eight planets.

THRUSTER – a small rocket engine used to change a spacecraft's direction

UNIVERSE – all of space and everything in it

2003–2010
Mars Exploration rover Spirit (landed 2004)

2003–present
Mars Exploration rover Opportunity (landed 2004)

2005–present
Mars Reconnaissance orbiter

2007–8
Phoenix Mars lander

2011–present
Mars Science Laboratory – Curiosity (landed 2012)

2013–present
MAVEN probe

2013–present
Mars Orbiter Mission

2016–present
ExoMars orbiter

A WALKER STUDIO BOOK